Research Report

Tankering Fuel on U.S. Air Transport Aircraft

An Assessment of Cost Savings

Tanguy Hubert, Christopher Guo, Christopher A. Mouton, James D. Powers

RAND Project AIR FORCE

Prepared for the United States Air Force
Approved for public release; distribution unlimited

RAND
CORPORATION

For more information on this publication, visit www.rand.org/t/RR837

Library of Congress Cataloging-in-Publication Data is available for this publication.

ISBN: 978-0-8330-8785-0

Published by the RAND Corporation, Santa Monica, Calif.

Preface

Faced with declining budgets, the United States Air Force (USAF) must reduce costs while retaining capability. Fuel expenditures represent a significant portion of operating and support costs, and the service has targeted fuel use in recent years through a series of initiatives. Air Mobility Command (AMC), the biggest fuel consumer within the USAF, asked RAND Project AIR FORCE (PAF) to examine a broad set of fuel saving initiatives to determine cost-effective options for reducing fuel use.[1] This report presents an analysis of tankering, which seeks to lower total fuel costs by carrying excess fuel when traveling from locations where jet fuel is less expensive than at the destination. We examine the savings potential of tankering for the C-5, C-17, and C-130 based on historical flying patterns, fuel price data, and a number of other factors. We also distinguish between tankering to provide fuel for the tankering aircraft on later flight legs versus tankering with the intention to offload extra fuel for use in other aircraft.

This analysis was part of the "Fuel Reduction for the Mobility Air Forces" project, commissioned by the AMC Director of Operations and conducted within the Resource Management Program of RAND Project AIR FORCE. It should be of interest to mobility air operations planners and those concerned with energy use within the Department of Defense.

RAND Project AIR FORCE

RAND Project AIR FORCE (PAF), a division of the RAND Corporation, is the U.S. Air Force's federally funded research and development center for studies and analyses. PAF provides the Air Force with independent analyses of policy alternatives affecting the development, employment, combat readiness, and support of current and future air, space, and cyber forces. Research is conducted in four programs: Force Modernization and Employment; Manpower, Personnel, and Training; Resource Management; and Strategy and Doctrine. The research reported here was prepared under contract FA7014-06-C-0001.

Additional information about PAF is available on our website:

http://www.rand.org/paf/

[1] PAF's assessment of a broad set of fuel saving initiatives is documented in Mouton et al. (2014a, 2014b).

Table of Contents

Figures

Tables

Summary

Aviation fuel use accounts for a large proportion of the U.S. Department of Defense (DoD) total petroleum use—about 50 percent according to a 2012 analysis by the Congressional Research Service. A worldwide increase in fuel prices from fiscal year (FY) 2005 to FY 2011 led to a 381 percent increase in DoD fuel spending. Understandably, DoD is keenly interested in actions that can reduce its fuel costs. Air Mobility Command (AMC), which has large fleets of aircraft providing airlift and refueling services to joint forces, uses about 28 percent of all DoD fuel and is therefore a natural target for reductions.

One cost-saving technique that has attracted the attention of policymakers is fuel tankering. Tankering involves carrying excess fuel on an aircraft (more than is required for the flight) when traveling from origins where jet fuel is less expensive than at the destination and is a common practice in commercial aviation. By shifting fuel purchases from expensive to cheaper locations, tankering can decrease the overall cost of fuel—even after accounting for the additional fuel burned flying at heavier-than-required weights. In recent overseas contingency operations in Afghanistan, tankering has been employed by the Tanker Airlift Control Center to generate significant cost savings by exploiting the large differences between fuel prices outside and inside the theater of operations, where they have been up to three times higher.

As part of a broader analysis of options to reduce fuel expenditures, AMC asked RAND Project AIR FORCE (PAF) to estimate the potential cost savings if AMC's airlift fleet were to tanker fuel to the maximum extent possible in peacetime, whenever it is cost-effective to do so. We compare the fuel costs of flights completed without tankering during FY 2012 with the estimated fuel costs of the same flights if they had tankered fuel. An important feature is the jet fuel procurement process within DoD: The Defense Logistics Agency (DLA) purchases fuel on behalf of DoD at fluctuating market rates from local suppliers and then sells fuel to customers such as AMC at fixed DLA standard prices. There are four different DLA standard prices, depending on location type. Therefore, we distinguish between tankering savings from the perspective of AMC based on prices charged by DLA and from the perspective of DoD based on market prices.

Using historical sortie data and independently gathered market price estimates, PAF simulated AMC decisions to tanker during peacetime operations based on either DLA standard prices or actual market prices. We address the changing potential for future cost savings as AMC transitions from wartime to peacetime operations. We also consider how the amount of information available on fuel prices and future planned missions affects expected savings in fuel costs. Finally, we examine how fuel offloading affects savings from tankering.

Key Findings

Cost Savings from Tankering

In our baseline/wartime scenario, tankering fuel would have saved AMC $151 million annually, about 2 percent of the total Air Force aviation fuel budget of $8.81 billion in FY 2012 (DLA, undated-c). Under this scenario, almost 24 percent of the sorties would carry excess fuel for cost-saving purposes, and almost 16 percent of the flights tankering fuel would be domestic flights. It is important to note that baseline scenario estimates were calculated from historical wartime sortie data, and past flying patterns may not fully reflect future flying patterns. The majority of savings generated by tankering fuel on C-5, C-17, and C-130 flights came from operations in Iraq and Afghanistan.

We find that significant savings can be achieved in peacetime (although the opportunities are fewer than in the recent contingency environment because of smaller price variations between operating locations). However, these savings would require changes to the way DLA shares price information and a compensation mechanism within DoD.

Sharing Price Information Between AMC and DLA

Currently, DLA does not provide market rate information to AMC. It sets different standard prices based on location and relationships with local suppliers (e.g., U.S. military installations and civilian airfields). This incomplete information creates a misalignment of incentives: From the perspective of AMC, tankering savings are based on differences within DLA standard prices, but from the perspective of DoD, savings are based on differences between market prices. As shown in Table S.1, we estimate that tankering decisions based on DLA standard prices currently available to AMC would save the command $8.6 million annually but would cause DLA to face an $11.9 million budget shortfall. The net result would be a loss of $3.3 million for DoD. On the other hand, if AMC made tankering decisions based on market rate information, DLA would see $56.5 million in annual savings, but AMC would incur a $31.1 million annual loss, resulting in $25.4 million net savings to DoD.

Table S.1. Tankering Savings for AMC and DoD in a Peacetime Scenario (in $ millions)

Level of Price Information	Savings for AMC	Savings for DLA	Net Savings for DoD
AMC tankers based on DLA standard prices (incomplete information)	8.6	–11.9	–3.3
AMC tankers based on actual market prices (complete information)	–31.1	56.5	25.4

Fuel Offloading

In the analysis described above, we assumed that each aircraft tankers "selfishly," never carrying more fuel than would be required for a potential next leg. We also examined the possibility of aircraft carrying extra fuel with the intention of offloading tankered fuel on the ground at the destination to refuel other aircraft. We found that offloading fuel increases the savings significantly. In particular, when AMC and DLA cooperate, offloading fuel from the tankering aircraft's fuel tank increases DoD annual savings by a factor of 3.0 during peacetime and 3.8 during wartime.

Conclusions and Recommendations

Our analyses led us to the following conclusions and recommendations:

AMC can take advantage of differences in DLA standard prices. Savings from tankering are possible even without access to market fuel price data, if AMC takes advantage of differences among DLA standard prices. These savings are simple to achieve, because the prices are known, but they may occur only in the short run as DLA has the option to make adjustments to its pricing schedule. In the long run, it makes sense for AMC and DLA to work together to develop a list of worldwide locations and the associated fuel costs.

We recommend that the USAF work with DLA to provide market price information to AMC to maximize the savings to DoD of peacetime tankering. Additionally, an internal compensation mechanism within DoD would be needed to shift some of the savings reaped by DLA to AMC to incentivize the command's participation. The need for a compensation mechanism would be eliminated if DLA simply charged market rates; however, this would introduce complications in the fuel budgeting process.

Tankering operations can be pursued to different extents and implemented in phases. A highly involved, complex tankering system might require optimization with extended visibility into future missions and aircraft assignments, real-time market prices for fuel at all locations, infrastructure investments to store and offload fuel, information technology investments to perform the linear programming techniques, and very high level coordination. However, as we have demonstrated, potential cost savings are possible even with less sophisticated tankering systems. Ideally, flight planners would be given a tool that they can use to input departure and arrival locations and mission payload, with a "lookup table" specifying the optimal fuel load. Development and trials of implementation could begin with basic tankering efforts. Given time and experience, a decision could be made on whether to invest further in more sophisticated implementations.

Acknowledgments

We are grateful for the support and feedback we received from the project sponsor, Gen Scott Goodwin, AMC/A3, and Maj Gen Scott Hanson.[2] Within Air Mobility Command (AMC), the Fuel Efficiency Office (FEO) was invaluable in our effort. Col Bobby Fowler (Chief of AMC/FEO) provided important vision and insights to our research. His efforts were crucial in defining this work. Col Keith Boone continued Col Fowler's lead, and his help as we concluded our effort was very much appreciated. Also from the FEO, Col Michael Horsey, Maj Darren Loftin, and Maj Daniel Ortwerth continually provided us with the contacts and data we needed for our work. Their enthusiasm for the topic was very much appreciated. The assistance of George Guirgis at UVair in providing timely price quotes of international jet fuel prices was invaluable. Several industry experts—Tom Kane, Ronald Lane, Richard Rolland, and John Dietrich at Atlas—gave us additional insights into commercial aviation fuel management. We are thankful for the input of several RAND colleagues: Michael Kennedy and David Orletsky provided expertise and advice throughout our research; Jim Dittus and Bobby Theologis helped us collect price information from airport fixed-base operators; Jerry Sollinger and Kristin Leuschner helped prepare this document; and Sean Bednarz, Khaled Abdelghany, and Thomas Light provided detailed technical reviews of this work.

[2] All ranks and offices were current as of the time of the research.

Abbreviations

AMC	Air Mobility Command
CRS	Congressional Research Service
DLA	Defense Logistics Agency
DFSP	Defense Fuel Supply Point
DoD	Department of Defense
FEO	Fuel Efficiency Office
FSII	fuel system icing inhibitor
FY	fiscal year
GDSS	Global Decision Support System
GPM	gallons per minute
ICAO	International Civil Aviation Organization
IPCIS	Into Plane Contract Information System
MAF	Mobility Air Forces
MAFPS	Mobility Air Force's Automated Flight Planning Service
MG	millions of gallons
MTOW	Maximum Takeoff Weight
NATO	North Atlantic Treaty Organization
PAF	Project AIR FORCE
USAF	United States Air Force

1. Introduction

Background

The Department of Defense (DoD) is the largest U.S. government consumer of energy, accounting for about 80 percent of the federal government's energy use. In particular, aviation fuel accounts for 50 percent of DoD total energy use according to a 2012 analysis by the Congressional Research Service (CRS). In recent years, DoD fuel costs have increased significantly; as fuel prices have gone up, DoD spending on petroleum rose 381 percent between fiscal year (FY) 2005 and FY 2011—even though its petroleum use decreased 4 percent over the same period (CRS, 2012). Understandably, DoD is keenly interested in measures that reduce its fuel use and energy costs. Air Mobility Command (AMC), with its large fleets of lift and tanker aircraft, uses about 28 percent of all DoD fuel and is therefore a natural target for reductions (Fritz, 2010).

Fuel costs can be reduced through energy efficiency and conservation measures. These measures include technology improvements (aerodynamics, aircraft weight, propulsion, etc.), and the optimization of fleet composition and flight and ground operations. A detailed analysis and comparison of these fuel savings alternatives can be found in a companion RAND publication (Mouton et al., 2014a, 2014b).

Another option for reducing aviation fuel costs is fuel tankering. Tankering is "the purchase of fuel in excess of that immediately required for the next flight leg, … [by] topping off the tanks at the cheaper stations to the extent the increased burn penalty and station supply allow" (Nash 1981). The "tankered fuel" is the amount of excess fuel carried during a flight. Although energy efficiency and conservation measures reduce fuel costs by decreasing fuel consumption, fuel tankering actually *increases* fuel consumption as a result of the degradation in cruise efficiency at higher weights. However, the practice *decreases* overall fuel cost by shifting some fuel purchases from expensive to cheaper locations.

Most commercial air carriers routinely use some sort of flight planning software to determine whether their flights should tanker fuel to reduce operating costs. Since the Arabian oil embargo of 1973, air carriers have developed least-cost fueling strategies for their flights (Darnell and Loflin, 1977; Nash, 1981; Stroup and Wollmer, 1992; Abdelghany, Abdelghany, and Raina, 2005; Kheraie and Mahmassani, 2012; Fregnani et al., 2013; Singh and Sharma, 2014). A recent study of major U.S. air carriers, including FedEx, United Parcel Service, and Continental Airlines, showed that fuel tankering generated up to $10 million per year in cost avoidance, depending on the air carrier considered (Lesinski, 2011).

Lesinski (2011) provides a recent study on potential tankering savings in a military context. Lesinski suggests that the U.S. Air Force (USAF) is missing out on significant cost savings through an exclusive emphasis on limiting total fuel consumption rather than also focusing on reducing the total cost of fuel. In recent overseas contingency operations in Afghanistan, tankering also has been employed by the Tanker Airlift Control Center to generate significant cost savings by exploiting the large differences between fuel prices outside and inside the theater of operations, where they have been up to three times higher (USAF, 2014).

Purpose of This Research

This report examines the option of tankering fuel on military transport aircraft as a cost-saving strategy for DoD, with a focus on AMC, which provides airlift and refueling services to joint forces. There is interest within AMC in understanding the potential cost benefits of using tankering. Mobility Air Force's Automated Flight Planning Service (MAFPS) currently combines several engineering inputs and information management factors to optimize mission fuel loads. The command might choose to expand on these inputs by incorporating data on fuel price and cost information (potentially from the Defense Logistics Agency [DLA]–Energy) to make decisions about tankering.

This analysis quantifies the potential savings by comparing the fuel costs of over 94,700 flights completed without tankering during FY 2012 with the estimated fuel costs of the same flights if they had tankered fuel to the extent possible. We also address the changing potential for future cost savings as AMC transitions from wartime to peacetime operations.

The work also examines other issues related to fuel tankering. First, we consider the relative benefits to the U.S. government as a whole (through DoD) compared to AMC. Next, we examine how fuel offloading affects the savings from tankering. Finally, we consider how the amount of information available, specifically on fuel prices and future planned missions, affects expected savings in fuel costs.

Approach

To understand the potential benefits from tankering, our analysis compares the fuel costs of historical AMC flights completed without tankering to the estimated fuel costs of the same flights. We build on Lesinski's prior work but use a more detailed methodology. We leverage a more extensive flight dataset, incorporate additional independently gathered price data, simulate the effects of informational barriers within DoD, and consider how future flying patterns might affect estimates of tankering savings. Our approach differs from that used in Lesinski's work in several important ways:

- Our research focuses exclusively on the C-5, C-17, and C-130 aircraft; KC-10 and KC-135 are beyond the scope of this report.[3]
- Lesinski extrapolated results from a two-week flight dataset limited to 2,029 relevant missions, but we conduct simulations using the complete FY 2012 mission set, including over 94,700 AMC flights with the potential for tankering fuel.
- Our analysis is not limited to the standard fuel prices set by the DLA, the fuel procurement agency of the U.S. Armed Forces: We also consider real market prices, including both spot and contract prices.

Our work also explores several new research directions not examined by Lesinski, including the value of tankering fuel in both peacetime and wartime scenarios, the role of cooperation in tankering decisionmaking, and how fuel offloading can affect savings from tankering.

Organization of the Report

The remainder of this report is organized as follows:

- Chapter 2 describes the flight and price datasets used in our research.
- Chapter 3 explains the four decision factors we considered in our analyses: capacity, fuel consumption penalty, fuel requirement of the next mission leg, and fuel price differential.
- Chapter 4 focuses on tankering as a fuel cost-saving approach and considers it from both the AMC and DoD perspectives.
- Chapter 5 examines the possibility of offloading all or part of the fuel tankered on AMC aircraft on arrival at the destination.
- Chapter 6 presents our conclusions and lays out some potential directions for future research.

[3] Our analysis excludes tankers because many sorties depart and arrive from the same location. In addition, fuel requirements for tanker aircraft are more complex than simply the amount of fuel required to fly a certain distance plus reserves because of offloading requirements.

2. Flight and Fuel Price Datasets

We quantify potential tankering savings by comparing the fuel costs of FY 2012 AMC flights completed without tankering to the estimated fuel costs of the same flights where fuel is tankered whenever it reduces total fuel cost. In this chapter, we briefly describe the data sources and research assumptions we employed. First, we describe the flight data used. Then, we describe the current jet fuel procurement process, including the DLA's process for setting standard fuel prices. Finally, we discuss our approach for estimating potential savings from tankering fuel.

Aircraft Availability and Flight Data

Flight data were obtained from the Global Decision Support System (GDSS), the Mobility Air Forces (MAF) information system. The dataset includes 122,921 sorties completed by C-5, C-17, and C-130 aircraft between October 1, 2011, and September 30, 2012. For each sortie, GDSS information includes the date, departure and arrival information, and payload.

For our analysis, we assume that differences in the performance characteristics of mission design series variants are small enough that a single representative variant can serve as the basis of our model with minimal effect on the analytical results. As an example, the C-5A, C-5B, C-5C, and C-5M are all considered to have the same C-5 characteristics.

We also assume that tankering fuel does not modify the operational capability and availability of AMC aircraft. In particular, we assume that payload is never decreased to tanker more fuel and that aircraft are never rerouted to increase tankering savings. We further assume that tankering fuel does not generate significant additional maintenance costs, since aircraft are operated within their structural limits.

Table 2.1 provides the aircraft characteristics assumed for this research. We further assume that the passengers' weight is negligible—an average of 6.8 passengers were on board across the 122,921 sorties considered.

Table 2.1. Characteristics of C-5, C-17, and C-130 Aircraft

	C-5	C-17	C-130
Maximum takeoff weight (lb)	840,000	585,000	175,000
Operating empty weight (lb)	400,000	282,500	83,628
Maximum usable fuel (lb)	332,500	244,854	43,560
Maximum cargo payload (lb)	285,000	164,900	47,812
Cruise speed (knots)	450	450	360

As seen in Table 2.2, the initial dataset was reduced by removing training flights that took off and landed at the same location (22 percent of the initial data). No tankering savings potential exists for these sorties, since the price of fuel is the same at both departure and arrival. The dataset was further reduced by excluding sorties that required refueling in the air (i.e., with mission fuel requirements that exceeded the maximum range according to payload). Finally, the departure or arrival locations of a small number of sorties (less than 1 percent) were unknown, and these sorties were removed.

This process left 94,726 sorties, which accessed 938 locations worldwide, to be examined for potential tankering savings (Figure 2.1). These sorties covered 11,122 routes, 167 of which accounted for 50 percent of the traffic.

Table 2.2. Number of Flights in the FY 2012 Flight Dataset

Description	C-5	C-17	C-130	Total	
Sorties in initial dataset	6,599	50,709	65,613	122,921	(100%)
Training flights (same departure and arrival)	1,475	6,180	19,747	27,402	(22%)
Excluded sorties[a]	25	151	506	682	(<1%)
Departure or arrival unknown	0	158	493	651	(<1%)
Sorties eligible for tankering	5,099	44,311	45,316	94,726	(77%)

[a]Excluded sorties were identified using approximate range-payload diagrams.

Jet Fuel Procurement Process and Standard Fuel Prices

The first step in assessing fuel tankering as a potential cost-saving strategy is to review the current jet fuel procurement process applicable to the C-5, C-17, and C-130 aircraft. AMC aircraft regularly refuel from local fuel suppliers at various locations worldwide: 1,420 different locations were included in the initial AMC dataset for FY 2012. Instead of paying suppliers directly, AMC pays DLA, DoD's fuel procurement agency, at standard rates that vary depending on the airfield. DLA pays the local suppliers at some market rate—either spot or contract rates. Figure 2.2 summarizes the steps in the procurement process.

DLA aims to use standard rates as a tool to remove day-to-day volatility and facilitate the planning and budgeting of jet fuel purchases. The purpose of DLA's standard fuel rates is to "insulate the Military Services from the normal ups and downs of the fuel marketplace" (DLA, undated-b). When DLA establishes the standard rates, it is intended that this will remain in place throughout the fiscal year. However, when actual fuel costs change significantly, DLA must update their standard rates in order to maintain solvency. In FY 2013 and FY 2014 there were no mid-year rate updates. As of April 2015, there has been one rate update for FY 2015.

Figure 2.1. Worldwide Locations of Flight Operations Eligible for Tankering Savings

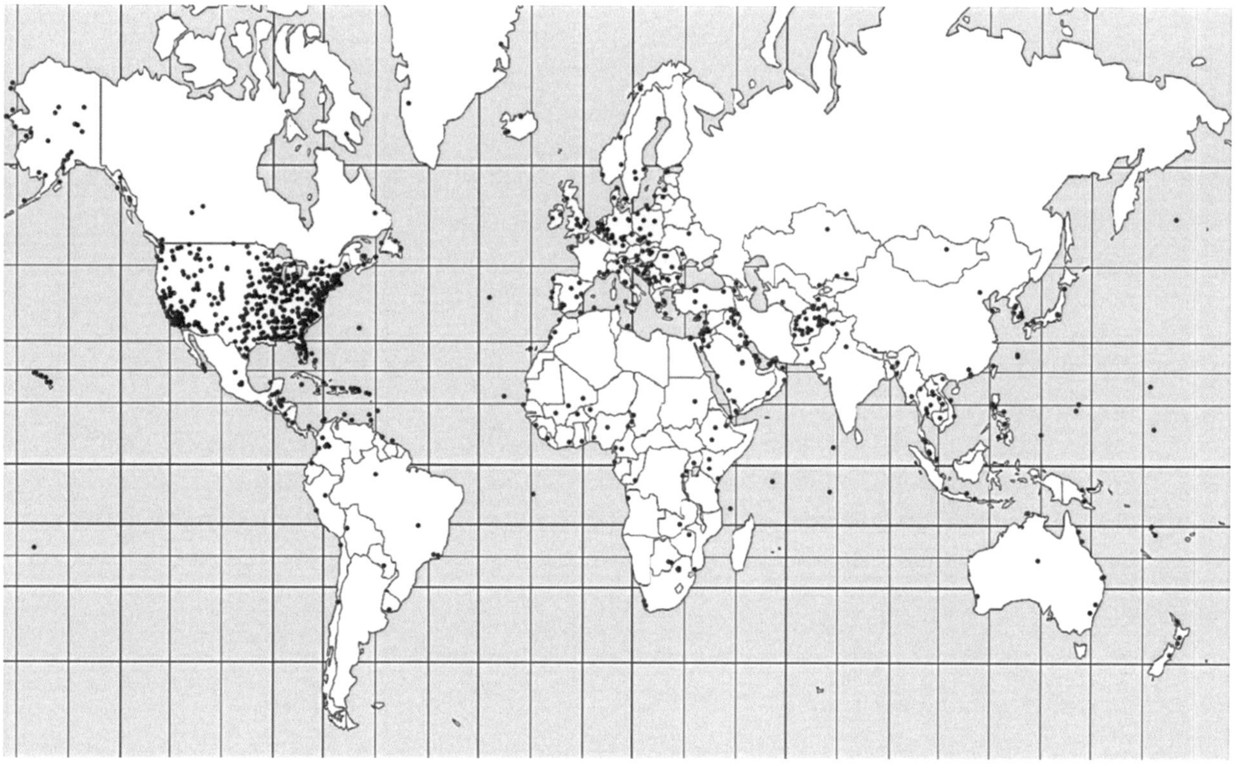

SOURCE: Authors.

Figure 2.2. Simplified Jet Fuel Procurement Process

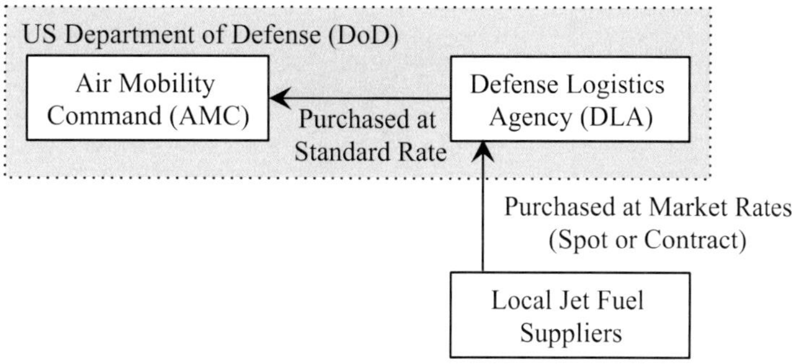

SOURCE: Authors.

DLA organizes the various airfields into four categories, each corresponding to a different *standard price*[4]:

[4] DoD and DLA reserve the term *standard price* for the standard rate common to all Defense Fuel Supply Point (DFSP) locations and use the term *escalated price* to refer to the standard rate at each of the three categories of non-DFSP locations. For simplicity, we do not make such distinction in this report and use the term *standard prices* for both DFSP and non-DFSP locations.

- The *Defense Fuel Supply Points* (DFSPs) are fueling stations at major U.S. military installations, including most stateside and overseas bases. DFSPs can be government-owned-and-operated, contractor-owned-and-operated, or government-owned-and-contractor-operated. We use the list of DFSPs as of January 1, 2011 (Lesinski, 2011).
- The *locations with contracts* are sites at which DLA has one or several ongoing fuel contracts with a local fuel supplier. Most of these contracts are "into-plane" contracts for Jet A-1 with fuel system icing inhibitor (FSII) or JP-8 fuel. The vendor information and latest escalated prices are available online from DLA's Into Plane Contract Information System (IPCIS) (DLA, undated-a).
- *Locations without contracts* are sites at which DLA is charged by the local fuel supplier at the current spot price.
- *Fields serviced by the North Atlantic Treaty Organization (NATO)* differ from the previous categories in that the standard fuel prices set by NATO reflect the fully burdened cost of providing fuel, including transportation, storage, security, and other costs. NATO prices are not DLA-generated; they are simply billed to the service that purchased fuel through DLA. We assume the same NATO-serviced fields as in Lesinski (2011, Appendix D).

Table 2.3 shows the distribution of the 1,420 airfields covered by the initial AMC dataset among the four DLA categories, with the corresponding standard fuel prices as of October 1, 2012. The table also indicates the location type for each of the sites covered by the 94,726 sorties that were eligible for tankering fuel during FY 2012. We use these standard fuel prices in our analysis.[5]

Table 2.3. Airfields in the AMC Dataset

Location Type	Distribution Among Initial AMC Dataset for FY 2012		Standard Fuel Price as of October 1, 2012 ($/gal.)
DFSP	220	(15.5%)	3.73
Location with contract	383	(27.0%)	4.26
Location without contract	814	(57.3%)	4.57
NATO-serviced fields	3	(0.2%)	9.00

Defining Savings from the Perspectives of AMC and DoD

In our analyses, we distinguish between tankering savings from the perspective of both AMC and DoD. To inform the former, we explain how AMC pays for its fuel and how the price for that fuel is set. For the latter, we outline our estimation methodology for global jet fuel market prices.

[5] Note that our analysis is not intended to evaluate the efficiency of the DLA standard price system.

We first evaluate the potential savings from tankering fuel on AMC flights from AMC's standpoint.[6] We assume that AMC is billed for its fuel purchases at the DLA standard prices—as is currently the case—and that DLA standard prices are the *only* prices that come into play when AMC decides whether to tanker fuel on a given mission. Under this first case, *fuel tankering is a cost-saving strategy if it decreases the total fuel cost paid by AMC to DLA*.

In a second case, we quantify the potential savings from tankering fuel on AMC flights from DoD's standpoint. This perspective is in many ways the more important one, since it reflects the true overall effect on the U.S. government. We continue to assume that DLA standard prices are the only ones that AMC uses to decide whether to tanker fuel on a given mission. In other words, we assume that DLA does not provide any information to AMC on the real market rates that could possibly guide AMC's decision. Under this scenario, *fuel tankering is a cost-saving strategy only if it decreases the total fuel cost paid by DoD—through DLA—to the local fuel suppliers*.

Quantifying the potential tankering savings from DoD's standpoint required that we estimate the real market rates for jet fuel at the various refueling locations considered. At 492 of these locations, DLA had at least one ongoing into-plane contract in place with a local fuel supplier as of May 31, 2013. For each location, we retrieved the corresponding escalated price from IPCIS. Depending on the products available locally, we retrieved the following, by order of priority: the price for JP-8 fuel, the price for Jet A-1 fuel with FSII, and the price for Jet A-1 fuel without FSII.

In addition to these 492 escalated contract prices, we obtained the into-plane spot prices for Jet A-1 fuel at 22 international locations with high AMC traffic (or close to locations with high AMC traffic) by contacting local fuel suppliers between May 20 and May 31, 2013. (A list of these locations with the corresponding prices appears in Appendix B.) We assume that DoD is not charged either sales or excise taxes at these locations or if it is charged, it can get reimbursed for these taxes similar to commercial air carriers flying internationally.

From these 22 spot prices and 492 escalated contract prices, we formed a pool of 514 "known" locations for which we have estimated the fuel market price. Among the 1,420 locations used by AMC flights during FY 2012, 465 (33 percent) are in the pool of known locations. We assigned to each of the remaining 955 locations the market price of the closest known location. Table 2.4 shows the cumulative percentage of locations covered as a function of the distance to the closest known location. As the distance from the closest known location and the percentage of locations covered increase, the probability also increases that the price assigned to the unknown location differs from the actual market price at that location.

To balance the increased coverage and associated reduced accuracy, *we calculate our final estimate of tankering savings for DoD by taking the average across four distinct flight subsets*. The results can be seen in Table 4.4. The first subset consists of the flights going through known

[6] Some AMC savings will be passed on to AMC customers through the Transportation Working Capital Fund.

Table 2.4. Cumulative Percentage of Fueling Locations Covered

Distance to Closest Known Location (mi)	0	<100	<200	<300
Percentage of locations covered (cumulative)	33	68	78	83

locations only. The second subset corresponds to the flights going through locations within 100 miles of known locations. The third and fourth subsets correspond, respectively, to flights going through locations within 200 and 300 miles of known locations. The first subset is the most precise in term of prices but covers only 33 percent of the locations. The fourth subset is potentially the most imprecise in term of prices but covers 83 percent of the locations.

3. Four Decision Factors to Consider When Tankering Fuel

In this chapter, we describe the four main factors that determine whether a given mission should tanker fuel for cost avoidance:

- the capacity available for carrying excess fuel
- the fuel burn penalty
- the amount of fuel required for the next mission leg
- the fuel price difference between the departure and arrival locations.

Capacity Available for Tankering Fuel

The capacity available for tankering fuel can be estimated as a function of the mission distance and payload. In the following, we assume initially that no additional internal fuel storage is used. The broader case where extra fuel can be stored in flexible containers placed in the cargo hold of the aircraft is discussed separately in Chapter 5.

The amount of fuel required for a given mission can be approximated as a nonlinear function of the mission distance and payload:

$$W_f(d, W_p) = a_1 + a_2 d + a_3 d^2 + a_4 W_p + a_5 {W_p}^2 + a_6 d W_p$$

W_f = amount of fuel required (lb)

d = mission distance (nautical miles)

W_p = mission payload (lb)

The coefficients a_i were derived from RAND-developed flight models for each of the three aircraft types considered[7] (Mouton et al., 2013; Bednarz et al., 2012; Rosello et al., 2009, 2011). Finally, for each sortie, d is calculated from the latitudes and longitudes of the departure and arrival locations using the haversine formula (that is, the mathematical formula for determining the shortest distance between two points on the earth).

The capacity available for tankering fuel can then be determined from the following parameters (Figure 3.1): (1) the amount of fuel $W_f(d, W_p)$ required for the mission (including 10 percent fuel reserves), (2) the tank capacity, and (3) the maximum takeoff weight (cf. Table 2.1).

---

[7] A 10 percent fuel reserve is factored into the model. An alternative to using a flight model is to use the Fuel Tracker database to empirically estimate the actual fuel burn relationship.

Figure 3.1. Aircraft Capacity for Tankering Fuel

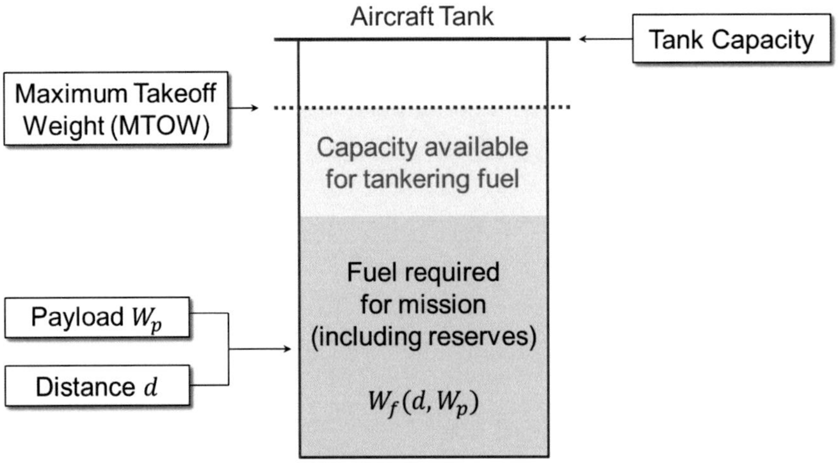

SOURCE: Authors.

Fuel Burn Penalty

Tankering fuel from one location to another increases takeoff weight and requires additional fuel to make up for the increased rate of fuel consumption. Fuel tankering is therefore a tradeoff between taking advantage of a favorable fuel price difference between two locations and consuming a fraction of this fuel in transit.

For a given aircraft type, on a given mission, to effectively tanker X lb of fuel to the arrival location, $(1+\xi)X$ lb must be loaded at departure, since ξX lb will be consumed along the way (Figure 3.2). We define ξ as the burn factor.[8] $K = (1+\xi)$ is sometimes referred to as the transport coefficient or transport factor (Airbus, 2004, pp.19–20):

$$K = \frac{\text{Excess Fuel Loaded at Departure}}{\text{Fuel Effectively Tankered to Destination}}$$

For a given mission, ξ can also be modeled as a function of the mission distance and payload. Tankering δW_f lb to the arrival location can be interpreted as increasing the payload W_p by δW_f and increasing the fuel W_f by $\xi \cdot \delta W_f$, which is the additional fuel consumed in transit. The average weight increase across the duration of the flight is $\delta W_f + \frac{\xi \cdot \delta W_f}{2}$, assuming a linear model for the fuel consumption.

[8] Lesinski (2011) provides typical ξ-values per flight-hour for the C-5, C-17, and C-130, although we use our own flight models to calculate fuel burn specific to the mission profile.

Figure 3.2. Fuel Use Between Departure and Arrival

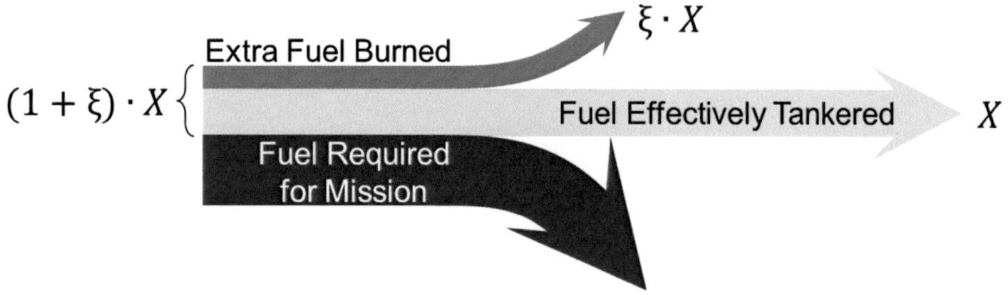

SOURCE: Authors.

Therefore, this relationship computes the change in fuel weight with respect to the change in the payload weight:

$$\frac{dW_f(d, W_p)}{dW_p} = \frac{\xi \cdot \delta W_f}{\delta W_f + \frac{\xi \cdot \delta W_f}{2}}$$

We can then solve for ξ and find that

$$\xi(d, W_p) = \frac{2(a_4 + 2a_5W_p + a_6d)}{2 - (a_4 + 2a_5W_p + a_6d)}$$

Fuel Required for the Next Mission Segment

In a typical flight (one that does not tanker fuel), the aircraft carries enough fuel to reach its destination and then refuels for the next leg. In a tankered flight, the aircraft carries excess fuel to be used in at least a portion the next leg(s). The savings result from buying less expensive fuel at the starting location rather than refueling with more expensive fuel at the transit point.

In the following, we assume that the fuel tankered from one location to another is never offloaded from the aircraft tank on arrival at destination. The feasibility and effect of fuel offloading are discussed separately in Chapter 5. From a cost-saving standpoint, we therefore avoid tankering more than the amount needed for the next mission legs, since tankering fuel always burns additional fuel.

In reality, most missions are multileg. For the aircraft types considered, flight schedules are typically known seven to ten days in advance on a rolling basis. Theoretically, the amount of fuel tankered can be optimized through use of linear programming techniques at a master planning level (Darnell and Loflin, 1977; Nash, 1981; Stroup and Wollmer, 1992; Abdelghany, Abdelghany, and Raina , 2005; Fregnani et al., 2013). However, in practice, flight planners do not necessarily have visibility over the complete flight schedule or know about the upcoming missions for that same aircraft beyond the most immediate one. From a methodological standpoint, the tail numbers are needed to track which aircraft completed which mission and to conduct an ex-post analysis using linear programming techniques, but these tail numbers are not always reported. For these reasons, we take a simplified approach and consider only the average profile of the next mission leg immediately following the current mission leg (Figure 3.3 right) rather than looking at the true multileg profile of each aircraft (Figure 3.3 left). We define an

Figure 3.3. True Multileg Mission Profile (Left) and Approximate Two-Leg Profile (Right)

SOURCE: Authors.

average profile for each of the 938 worldwide locations considered. For a given location, the average profile is defined as the average distance and average payload across all the sorties that departed from that location during FY 2012. This average profile is then used as the profile of the next mission leg for any mission flying into that location.

This approach reduces the amount of information needed and simplifies the decision process to solving a simple equation. It is not simply an analytical construct but also can be used practically as part of a future decision tool for the USAF.

Fuel Price Difference

Fuel tankering is driven by the differences in the price of aviation fuel between various locations. Consider an aircraft flying from A to B, with π_A being the fuel price at A and π_B being the fuel price at B (Figure 3.4). Tankering fuel from A to B can generate savings if the value of one unit of fuel at B is more than the value of one unit of fuel at A plus the cost to tanker one unit of fuel from A to B—the fuel burn penalty. Therefore, if $\pi_B - (1 + \xi) \cdot \pi_A > 0$, then the aircraft should tanker fuel from A to B within the tankering capacity available.

Algorithm 1 describes in pseudo-code a decision algorithm, factoring the four decision factors discussed in this chapter, to determine whether a given mission should tanker fuel.

Figure 3.4. Fuel Tankering Parameters

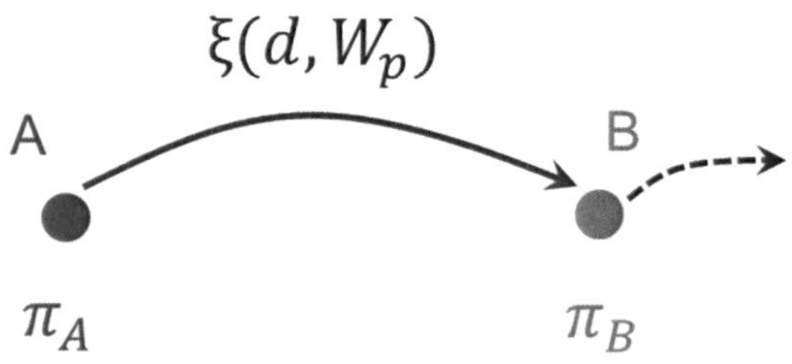

SOURCE: Authors.

Algorithm 1. Decisionmaking Process to Decide Whether and How Much to Tanker

if tankering capacity > 0

then calculate $\xi(d, W_p)$; calculate next-leg fuel requirement

    if $\pi_B - (1 + \xi) \cdot \pi_A > 0$

    then tanker fuel within tankering capacity and next-leg fuel requirement

    end if

end if

4. Is Tankering Fuel on AMC Aircraft a Cost-Saving Strategy?

This chapter considers savings that could be generated by tankering from two perspectives—first that of AMC and then that of DoD. For AMC, the value of tankering depends on relative differences in DLA standard prices; for DoD, tankering value depends on the actual market price.[9] The final section of the chapter discusses the effects of information-sharing between DLA and AMC on cost savings.

As will be discussed in more detail below, we find that it is possible for AMC to save money by tankering, whereas DoD incurs increased net costs if AMC uses the technique. In terms of overall savings for the U.S. government, the DoD perspective is the more important one. This difference can occur because AMC may not have adequate information on which to base its tankering decisions.

The AMC Perspective

Baseline Scenario

Results show that in the baseline scenario, tankering fuel could save AMC $150 million—about 2 percent of the total Air Force aviation fuel budget of $8.81 billion in FY 2012 (DLA, undated-c). Almost 24 percent of the sorties would carry excess fuel for cost-saving purposes, and almost 16 percent of the flights tankering fuel would be domestic flights. Table 4.1 presents results for each aircraft type.

It is important to realize that the value of tankering fuel from AMC's standpoint fluctuates over time and across space. The value of tankering at a given point in time depends on the

Table 4.1. Tankering Savings for AMC Under the Baseline Scenario

Aircraft Type	Annual AMC Savings ($M)	Total Tankered (MG)	Extra Fuel Burned (MG)	% of Flights Tankering Fuel	% of Domestic Flights Among Tankering Flights[a]
C-5	9.60	3.90	0.71	7.45	17.89
C-17	118.23	41.99	4.20	18.48	18.93
C-130	22.75	11.61	0.63	30.66	13.66
Total	**150.58**	**57.50**	**5.54**	**23.71**	**15.66**

NOTE: MG = millions of gallons.
[a]Domestic flights are defined as departing from and arriving at locations in contiguous U.S. states, Alaska, and Hawaii. (Scenario includes FY 2012 flights eligible for tankering and DLA standard prices as of October 1st, 2012.)

[9] A third perspective is that of DLA, which buys at market prices and sells to AMC at standard prices.

relative differences in standard prices, which DLA updates several times a year. The value of one unit of tankered fuel varies across routes because of geographic differences in fuel prices.

To illustrate the fluctuations in potential tankering savings across time, we re-ran the simulations for the 12 successive lists of standard fuel prices published by DLA between April 2009 and October 2012 (Figure 4.1 and Appendix A). We assumed that the flight schedule remains unchanged (94,726 sorties were eligible for tankering fuel during FY 2012). Results show that the annual tankering savings for AMC range from $95 million (when using the September 2009 price list) to $150 million (when using the October 2012 price list).

We also quantified the spatial fluctuations by tracking AMC potential savings per location. This analysis revealed that 95 percent of AMC savings under the initial baseline scenario come from flights going through Iraq and Afghanistan. In particular, the three expensive NATO

Figure 4.1. Temporal Evolution of Potential Tankering Savings for AMC

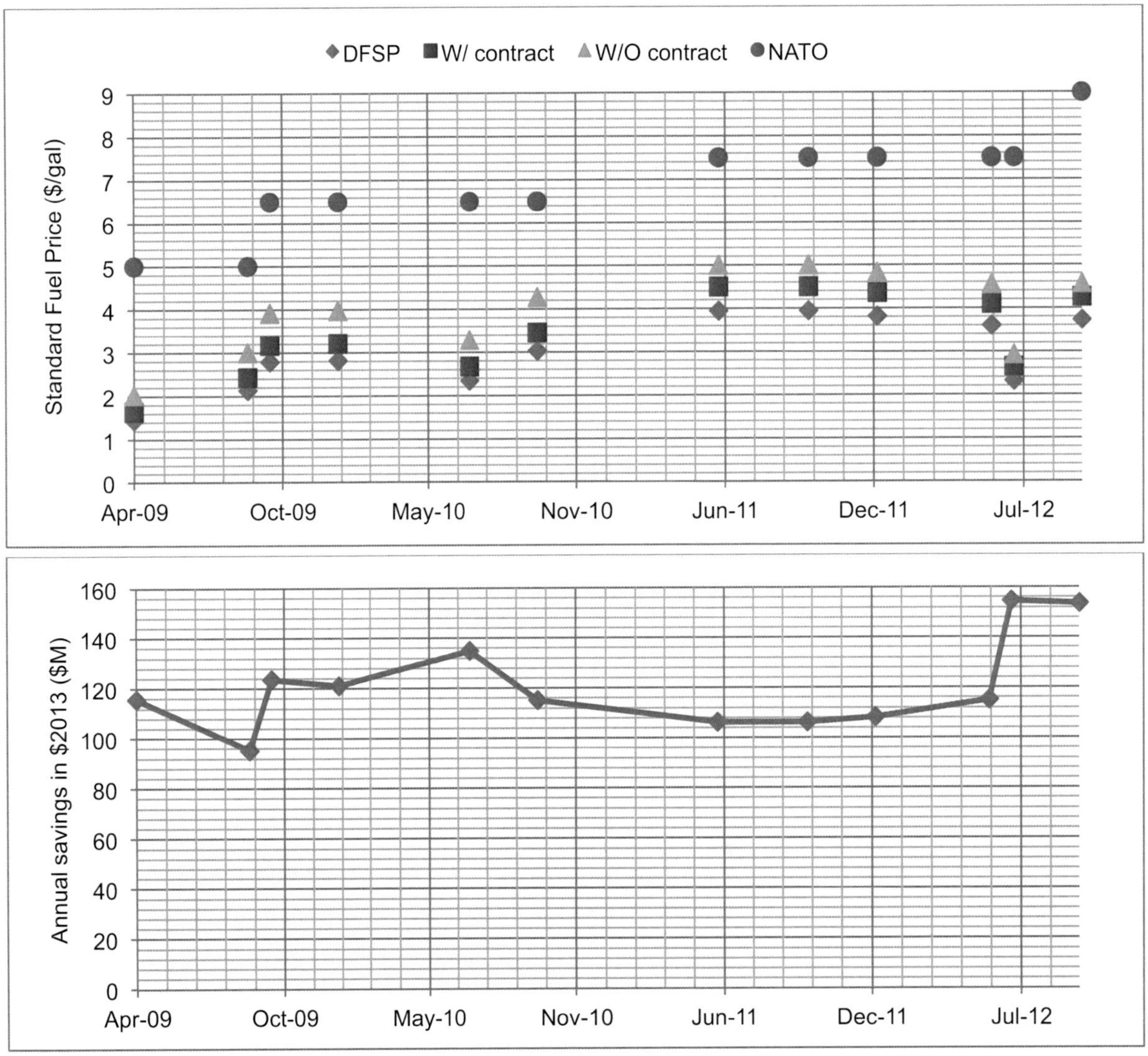

locations in Afghanistan—Camp Bastion, Kandahar, and Kabul—would attract 50 percent of the total tankered fuel quantity over the course of the year.

Wartime Versus Peacetime Scenarios

This concentration of potential tankering savings in theaters of operation led us to redefine our initial baseline scenario as the "wartime scenario" and to build another scenario defined as the "peacetime scenario" for comparison purposes.

To build the peacetime scenario, we assume that no flights operate into or out of Iraq and Afghanistan. We further assume that the peacetime activity can be reasonably modeled by selecting from the wartime scenario all the flights that do *not* go through Iraq and Afghanistan. The amounts of fuel that could potentially be tankered on these selected flights as well as the corresponding savings generated are then scaled by aircraft type to the total jet fuel consumption assumed in the FY 2014 Air Force President's Budget for the C-5, C-17, and C-130 aircraft considered (peacetime).

A comparison of AMC tankering savings under the wartime and peacetime scenarios appears in Table 4.2. Simulation results show that, although savings are still available, the annual AMC savings decline from $150 million under the wartime scenario to only $8.6 million under the peacetime scenario.

Thus, we conclude for this first case that tankering fuel is a cost-saving strategy for AMC, but that the vast majority of the savings come from contingency operations.

Table 4.2. Comparison of AMC Tankering Savings for Wartime and Peacetime Operations

Scenario	Annual AMC Savings ($M)	Total Tankered (MG)	Extra Fuel Burned (MG)	% of Flights Tankering
Wartime	**150.58**	57.50	5.54	23.71
Peacetime	**8.60**	23.90	1.90	4.00

The DoD Perspective

From DoD's standpoint, the potential savings from tankering fuel on AMC flights are calculated for both wartime and peacetime scenarios, as defined in the previous section. Results appear in Table 4.3, and Table 4.4 provides a more detailed breakdown of DoD savings.

Under the wartime scenario, unilateral fuel tankering on AMC flights also benefits DoD by generating annual savings close to $79 million. However, during peacetime, simulations show that unilateral tankering on AMC flights actually costs DoD an additional $3.3 million.

During wartime, the DLA standard price information appears to be sufficient to enable AMC to generate savings for both AMC and DoD (i.e., for the U.S. taxpayers). In particular, standard

Table 4.3. Tankering Savings for AMC and DoD (Without AMC-DLA Cooperation) ($ millions)

Scenario	Savings for AMC	Net Savings for DoD
Wartime	150.6	78.9[a]
Peacetime	8.6	–3.3[a]

[a]Refer to Table 4.4 for a detailed breakdown of savings.

Table 4.4. Breakdown of Tankering Savings by Aircraft Type and Flight (Without AMC-DLA Cooperation) ($ millions)

Scenario	Flight Subset (Distance to Closest Location, mi)	Net Savings for DoD			
		C-5	C-17	C-130	Total
Wartime	0	0.8	39.7	–1.9	38.7
	<100	2.1	90.3	4.8	97.1
	<200	1.9	84.4	4.1	90.3
	<300	1.8	83.7	4.0	89.5
	Final estimate (average across subsets)	1.6	74.5	2.7	**78.9**[a]
Peacetime	0	–0.3	0.3	–1.6	–1.7
	<100	–0.5	–2.1	–1.0	–3.6
	<200	–0.4	–2.2	–1.3	–3.9
	<300	–0.5	–2.4	–1.3	–4.2
	Final estimate (average across subsets)	–0.4	–1.6	–1.3	**–3.3**[a]

[a]Final values are reported in Table 4.5.

prices correctly signal that Afghanistan is the most expensive area to refuel from by assigning to almost every Afghan airfield one of the two most expensive standard prices categories: location without contract or a NATO-serviced location. Large savings are generated by the magnitude of the market price difference between Afghanistan and the rest of the world and the large quantities of fuel tankered to Afghanistan.

However, during peacetime, the standard price information is not sufficient to signal the true relative fuel price differences that exist between the various refueling locations. This situation would lead AMC to conclude that tankering is cost-saving on certain routes, based on DLA standard prices information, when an analysis using the real market prices leads to the opposite conclusion. For example, an AMC aircraft flying from Ramstein (ETAR) to Kalamata (LGKL) is likely to tanker fuel based on the DLA standard prices as of October 2012, since fuel is cheaper at ETAR ($3.73 per gallon at ETAR, $4.57 per gallon at LGKL). However, the corresponding market prices ($4.44 per gallon at ETAR, $3.03 per gallon at LGKL) show that, in reality, fuel is cheaper at LGKL. Thus, tankering fuel on this route incurs losses to the DoD.

Thus, we conclude for this second case that tankering fuel unilaterally on AMC flights is a cost-saving strategy for both AMC and DoD under the wartime scenario, even when AMC decisions are based only on DLA standard prices (incomplete information). However, under the peacetime scenario, tankering fuel unilaterally on AMC flights with incomplete information incurs losses to the DoD.

What If AMC and DLA Cooperate?

In the previous sections, we quantified the potential savings for both AMC and DoD under two scenarios—wartime and peacetime—assuming *incomplete information*, i.e., AMC does not know the true market rates. In this section, we examine how cooperation between AMC and DLA could affect both AMC and DoD savings.

We define cooperation as DLA providing real market rate information to AMC and by AMC using these market rates—instead of the standard rates to guide tankering decisions. We subsequently refer to this arrangement as *complete information*. We continue to assume that AMC is charged at the DLA standard rates. In other words, real market rates are not used by DLA for billing purposes but simply as a way to guide AMC decisionmaking more effectively.[10]

Table 4.5 compares the simulation results obtained for the wartime and peacetime scenarios assuming either incomplete or complete information. Under the wartime scenario, simulation results show that cooperation between AMC and DLA leads to a 40 percent increase in DoD savings (with almost $110.2 million saved annually) whereas AMC savings remain roughly constant. In wartime, DLA loses money, even when AMC tankers based on market prices, because of locations where its standard price is below market price.

Under the peacetime scenario, cooperation between AMC and DLA corrects the signaling issues previously observed when using DLA standard prices to guide AMC decisionmaking. When AMC and DLA cooperate, DoD now saves $25 million annually instead of losing $3.3 million. However, AMC now loses more than $31 million annually. In other words, under the peacetime scenario with complete information, AMC effectively gets penalized for tankering fuel. For example, an AMC aircraft flying from Rome (LIRF) to Ramstein (ETAR) is likely to tanker fuel based on the real market prices, since fuel is cheaper at LIRF ($3.10 per gallon at LIRF, $4.44 per gallon at ETAR). However, the corresponding DLA prices as of October 2012 ($4.57 per gallon at LIRF, $3.73 per gallon at ETAR) show that the cost charged to AMC at

---

[10] Our estimates of savings, which originate from differences in market prices, are subject to uncertainty because a complete set of market prices was not known.

Table 4.5. Tankering Savings for AMC and DoD: Comparison of Complete and Incomplete Information ($ millions)

Scenario	Level of Price Information	Savings for AMC	Savings for DLA	Net Savings for DoD
Wartime	AMC tankers based on DLA standard prices (incomplete information)	150.6	–71.7	78.9
	AMC tankers based on actual market prices (complete information)	147.2	–37.0	110.2
Peacetime	AMC tankers based on DLA standard prices (incomplete information)	8.6	–11.9	–3.3
	AMC tankers based on actual market prices (complete information)	–31.1	56.5	25.4

LIRF is actually higher than at ETAR. Tankering fuel on this route therefore incurs losses to AMC. *In conclusion, cooperation on fuel tankering decisions between AMC and DLA always leads to increased savings for DoD and therefore the U.S. government as a whole. However, results under the peacetime scenario with complete information show that some compensation mechanism may be required within DoD to maintain an incentive for AMC to tanker fuel.*

5. Fuel Offloading: Economic Effect and Feasibility

In this chapter, we examine the possibility of offloading all or part of the fuel tankered on AMC aircraft on arrival at the destination to quantify how fuel offloading affects the tankering savings. We also offer a preliminary assessment of the potential barriers to implementation that may serve as a basis for future analysis.

Assumptions Used in Fuel Offloading Analysis

In previous analyses, we assumed that the fuel tankered from one location to another was never offloaded from the aircraft on arrival at destination. This led us to limit the amounts of fuel tankered based on the fuel requirements for the next mission leg.

In this analysis, we now assume that all or part of the fuel tankered may be offloaded on arrival at the destination if this leads to increased tankering savings. We also assume, based on past observed flight activity, that there is a consistent demand for jet fuel at each of the worldwide locations considered. In other words, demand is not a limiting factor, since it is unlikely that the amount of fuel available for offload would exceed the amount of fuel consumed at any given location. Specifically, this is because the outbound traffic from any airfield generally requires far more fuel than AMC would tanker in.

We further assume that the fuel offloaded can be temporarily stored in fuel trucks before being reloaded into other aircraft, either immediately or within a few hours. Instead of refilling at the tank farm, these fuel trucks simply refill from an aircraft that has just landed while the regular cargo payload is being offloaded. No additional stand-alone storage is therefore needed. We do not consider aircraft-to-aircraft fuel transfers because of their operational complexity.

Finally, carrying excess fuel in the aircraft's fuel tanks is the most obvious way to tanker fuel. But additional savings might also be generated by storing excess fuel in flexible "pillow tanks" placed in the aircraft cargo hold when space is available. On arrival at the destination, these pillow tanks would be defueled, folded, and stored in the aircraft for the next rotation.

Economic Effect on Tankering Savings

We quantified the effect of fuel offloading across multiple cases modeling different configurations: wartime or peacetime, complete or incomplete information, and offloading only from the aircraft tank or from both the tank and some flexible containers. Several observations can be made based on the results obtained (Table 5.1).

First, offloading fuel increases the savings significantly. In particular, when AMC and DLA cooperate, offloading fuel from the tank increases DoD annual savings by a factor of 3.0 during peacetime and 3.8 during wartime, compared to the no-offloading baseline. In the wartime case,

Table 5.1. Savings from Offloading Fuel Under Multiple Scenarios ($ millions)

		Wartime Scenario		Peacetime Scenario	
Level of Price Information	**Offloading of Tankered Fuel**	**Savings for AMC**	**Savings for DoD**	**Savings for AMC**	**Savings for DoD**
AMC tankers based on DLA standard prices (incomplete information)	No offloading	150.6	78.9	8.6	–3.3
	Offloading from tank only	650.0	187.6	121.0	–19.3
	Offloading from tank and cargo hold	1,010.2	244.1	182.0	–29.9
AMC tankers based on actual market prices (complete information)	No offloading	147.2	110.2	–31.1	25.4
	Offloading from tank only	383.9	422.3	–68.9	76.4
	Offloading from tank and cargo hold	536.5	614.7	–103.0	114.7

the $422 million in annual savings obtained represents almost 6 percent of the FY 2012 Air Force aviation fuel budget (DLA, undated-c). Adding the ability to tanker fuel in flexible containers increases the savings by factors of 4.5 during peacetime and 5.6 during wartime.

In addition, offloading fuel reinforces a number of patterns previously observed in the non-offloading case. From DoD's standpoint, cooperation between DLA and AMC is always desirable. Cooperation increases DoD's savings during wartime and prevents DoD from incurring losses during peacetime. From AMC's standpoint, under incomplete information, the ability to offload fuel increases the savings resulting from tankering fuel on AMC aircraft. However, under complete information, AMC gets penalized even more than in the non-offloading case. This reemphasizes the need for some compensation mechanism within DoD to maintain an incentive for AMC to tanker fuel to generate savings for the DoD as a whole.

Feasibility of Fuel Offloading

Offloading all or part of the fuel tankered on AMC aircraft appears to increase DoD's savings significantly. However, as noted above, the need for AMC's participation and "buy-in" represents one potential barrier to implementation that would need to be addressed.

In this section, we identify a list of other potential barriers to implementation and offer a preliminary assessment of their relevance.

Time

One concern is that time spent on defueling operations may slow down the rotations of AMC aircraft. Fuel can be offloaded from an aircraft using either of two methods: suction defueling, which uses suction applied at the aircraft's ground refueling adapter, or pressure defueling, which uses onboard pumps to pump the fuel off the aircraft (Langton et al., 2009, p. 58). Tankered fuel can also potentially be defueled from flexible containers stored in the cargo hold.

The same fuel trucks are used for refueling and defueling operations. These trucks include the R-11 tank truck (the primary mobile refueling vehicle for the USAF), the older R-9 truck, and commercial trucks such as the Rampmaster 17.5K (USAF, 2010c).

Ground refueling of large transport aircraft can be performed at rates up to 800–1,000 gallons per minute (GPM) whereas defueling operations are usually slower, in the 150–300 GPM range (Table 5.2). The C-17 is unusual in that it can deliver fuel through either one or both of its single point receptacles at up to 520 GPM, depending on the number of internal booster pumps used (U.S. Army, 1998, 2006).

For each aircraft type, we compared the times needed to offload the average amount of fuel tankered per flight—assuming the conservative case in which all of the tankered fuel was to be offloaded—to the planning times used to estimate the time needed to offload the regular mission payload (Table 5.3). Current Air Force guidelines indeed authorize winching cargo or movement of nonpalletized self-propelled vehicles or equipment into or out of aircraft in conjunction with concurrent servicing for the C-5, C-17, and C-130 aircraft (USAF, 2013, p. 5-5).

Results show that even at a defueling rate of 175 GPM, defueling operations would never exceed the planning times for offloading the regular mission payload. *Therefore, provided that a sufficient number of trucks are available, we conclude that defueling operations are unlikely to slow down AMC aircraft rotations based on standard planning times.*

Table 5.2. Fuel Truck Characteristics

Truck Type	Capacity (gal)	Capacity (lb)	Refueling Rate (GPM)	Defueling Rate (GPM)
R-9[a]	5,000	33,550	600	200
R-11[a]	6,000	40,260	600	150–300 (USAF guidelines: 175[b])
Rampmaster 17.5K[c]	17,500	117,425	800	—
Titan Aviation 40,000L[d]	10,600	71,126	1,000	100 (suction alone) 300 (suction + pressure)

[a]Health, 2005, p. 14.
[b]USAF, 2012.
[c]Rampmaster, undated.
[d]Titan Aviation, undated.

Table 5.3. Fuel Offloading Times

	Average Fuel Tankered per Flight (gal) Wartime Scenario, Incomplete Information		**Fuel Offloading Times at 175GPM (min)**		**Fuel Offloading Times at 300GPM (min)**		**Passenger and Cargo Operations Wartime Planning Times (hrs + min)**	
Aircraft Type	Offloading from tank only	Offloading from tank and cargo hold	Offloading from tank only	Offloading from tank and cargo hold	Offloading from tank only	Offloading from tank and cargo hold	Offload	Onload
C-5	28,288	41,890	2+41	3+59	1+34	2+20	4+15	4+15
C-17	24,844	32,704	2+22	3+07	1+23	1+49	3+15	3+15
C-130	4,729	11,540	0+27	1+06	0+16	0+38	2+15	2+15

Manpower Costs

A second concern is that defueling operations may generate additional manpower costs because of relatively slow defueling rates. Defueling rates range from 150 to 300 GPM (Table 5.2). In contrast, a Navy manual used for facility planning assumes that trucks can accept fuel from the refueling facility at 450 to 600 GPM (U.S. Navy, 2005). In other words, a truck would refuel 1.5 to 3.5 times more slowly from an aircraft than from a tank farm.

Assuming that it normally takes about the same time for the truck first to refuel from the tank farm and then to transfer fuel into the aircraft tank, fuel offloading could potentially slow down the whole refueling process by a factor of 1.25 to 2.25. Manpower costs would therefore increase accordingly.

To estimate the share of manpower costs in the final market price, we reviewed the DLA contract prices at the 38 locations where DLA has both into-plane and into-truck contracts as of September 1, 2013. At 31 of these 38 locations, the into-plane and into-truck contract prices were identical; at the remaining locations, the spread ranged from 1 to 40 cents per gallon.

We conclude that fuel offloading would probably increase manpower costs, but this increase would be small compared to the total fuel costs.

Fuel Quality

A third concern is that defueling operations may contaminate the fuel. Prevention and testing procedures already exist to prevent contamination when refueling aircraft (whether commercial or military) (see, for instance, Federal Aviation Administration, 1976, 1978, 1985). Existing fuel trucks are equipped with appropriate filters and water separators with GPM rates that match the truck fueling and defueling rates (Health, 2005, p. 14).

Since the fuel trucks currently in use already have the required pumping, metering, and filtering capabilities, it is likely possible to transfer tankered fuel from the aircraft tank or from flexible containers into fuel trucks without contaminating the fuel, possibly using formalized defueling procedures. Then the fuel truck could be used to refuel other aircraft following the existing refueling procedures.

Wet wing defueling—the transfer of fuel from fixed-wing aircraft fuel tanks to collapsible fabric tanks or tank semi-trailers—is already performed in forward areas of operations (USAF, 2013). Air Force pamphlet 23-221 states that "using the correct procedures, wet wing defueling from the single point refueling port of [approved aircraft (including the C-5, C-17 and C-130)] into collapsible fabric tanks/bladders or tank semi-trailers can be done with an acceptable degree of risk" (USAF, 2006).

In conclusion, defueling operations present risks of fuel contamination, similar to refueling operations. However, the pumping, metering, and filtering equipment required for conducting defueling operations is already deployed on the ground, and wet wing defueling operations are already performed in forward areas of operations. *Therefore, if appropriate procedures are formalized and followed, defueling operations could most likely be performed with the same degree of risk as refueling operations.*

Ground Logistics

Logistics services on the ground constitute the fourth and probably most significant potential barrier that we identified.

The concern is twofold. First, defueling operations would make the truck scheduling problem more complex. We assumed that during defueling operations, fuel trucks refill from an aircraft that has just landed instead of refilling at the tank farm. Defueling operations would therefore increase the traffic of fuel trucks at the apron (leading to possible safety concerns) with truck drivers going from one aircraft to another and refilling from multiple locations.

Second, defueling operations could possibly lead to a shortage in refueling equipment. We previously assumed that tankered fuel was always offloaded into trucks and stored temporarily in these trucks before being reloaded into other aircraft, either immediately or within a few hours. Table 5.4 shows the equivalent number of trucks—either R-11 or Rampmaster 17.5K—required to temporarily store the average amount of fuel tankered per flight under the wartime scenario with incomplete information. The corresponding number of trucks may not be available locally when fuel has to be stored for a few hours before being reloaded. And when available, their use as temporary fuel containers for a few hours may interfere with other fueling operations.

We conclude that ground logistics may constitute a significant barrier to the implementation of fuel offloading. Additional analysis is needed to (1) determine whether the existing truck fleets at the various locations considered could handle increased use resulting from defueling operations, (2) determine whether expanding the fleet at certain locations could be cost-beneficial, and (3) confirm our assumption that demand at each location is not a limiting factor. We believe in general that the outbound requirement for fuel at any airfield would generally always exceed the amount of fuel tankered in by AMC aircraft; however, important exceptions may exist. These topics may be an area for future research.

Table 5.4. Average Fuel Tankered per Flight

	Average Fuel Tankered per Flight (gallons) Wartime Scenario, Incomplete Information		**Equivalent number of R-11 needed**		**Equivalent number of Rampmaster 17.5K needed**	
Aircraft Type	Offloading from tank only	Offloading from tank and cargo hold	Offloading from tank only	Offloading rom tank and cargo hold	Offloading from tank only	Offloading from tank and cargo hold
C-5	28,288	41,890	5	7	2	3
C-17	24,844	32,704	5	6	2	2
C-130	4,729	11,540	1	2	1	1

Summary of Barriers

Table 5.5 lists the four potential barriers and summarizes our conclusions.

Table 5.5. Potential Barriers to Implementation of Fuel Offloading

Barrier	Concern	Relevance	Comments
Time	Defueling operations could slow down rotations of AMC aircraft	○	Existing fuel trucks can defuel within time required to offload the regular cargo payload
Manpower costs	Defueling operations could increase manpower costs	◐	The increase in manpower costs would probably be negligible compared to the total fuel costs
Fuel quality	Defueling operations could contaminate fuel (water, rust, sand, etc.)	◐	Existing trucks are already equipped with filters and separators; prevention and testing procedures already exist
Ground logistics	Defueling operations could lead to equipment shortage	●	Holding the tankered fuel in trucks could interfere with other fueling operations

6. Conclusions and Recommendations

This report examined the option of tankering fuel on military transport aircraft as a cost-saving strategy for DoD. We compared the fuel costs for more than 94,700 C-5, C-17, and C-130 flights completed without tankering during FY 2012 to the fuel costs of the same flights with tankering. We simulated multiple cases and modeled different configurations, including wartime or peacetime, different levels of cooperative data-sharing within DoD, and the possibility of offloading tankered fuel.

In this chapter, we will summarize the key findings from our model and then will offer an interpretation and guidance for AMC.

Our analysis revealed several key findings:

- Most savings generated by tankering fuel on C-5, C-17, and C-130 flights would come from in-theater wartime operations. DoD can save up to $110 million annually under the wartime scenario compared to $25 million under the peacetime scenario (Figure 6.1).
- From DoD's standpoint, sharing information about the true market rates of fuel between DLA and AMC is always preferable (Table 4.5). Cooperation on fuel tankering decisions always leads to increased savings for DoD and therefore the U.S. government as a whole.
- Under the wartime scenario, unilateral fuel tankering on AMC flights generates savings for both AMC and DoD, regardless of whether information is shared between AMC and DLA (Table 4.5). Under the peacetime scenario (which excludes flights to and from Afghanistan and Iraq), with incomplete information, it is possible for DoD to incur losses from AMC tankering activity, as AMC makes tankering decisions based on DLA standard prices (i.e., not knowing the true market prices for fuel). With complete information, AMC itself may actually incur losses while DoD as a whole experiences savings as a result of tankering (Table 4.5). This is because AMC makes tankering decisions based on actual market prices but is charged DLA standard prices. Thus, some compensation to AMC may be required within DoD to maintain an incentive for AMC to tanker fuel.
- Fuel offloading—from the aircraft tank or flexible containers stored in the aircraft cargo hold—can increase DoD tankering savings by up to 460 percent (Table 5.1).
- Our findings reveal that the cost savings of tankering vary greatly depending on the scenario (wartime versus peacetime, complete versus incomplete information, AMC versus DoD standpoint, capability of offloading tankered fuel).

Figure 6.1. Fuel Tankering by USAF Transport Aircraft

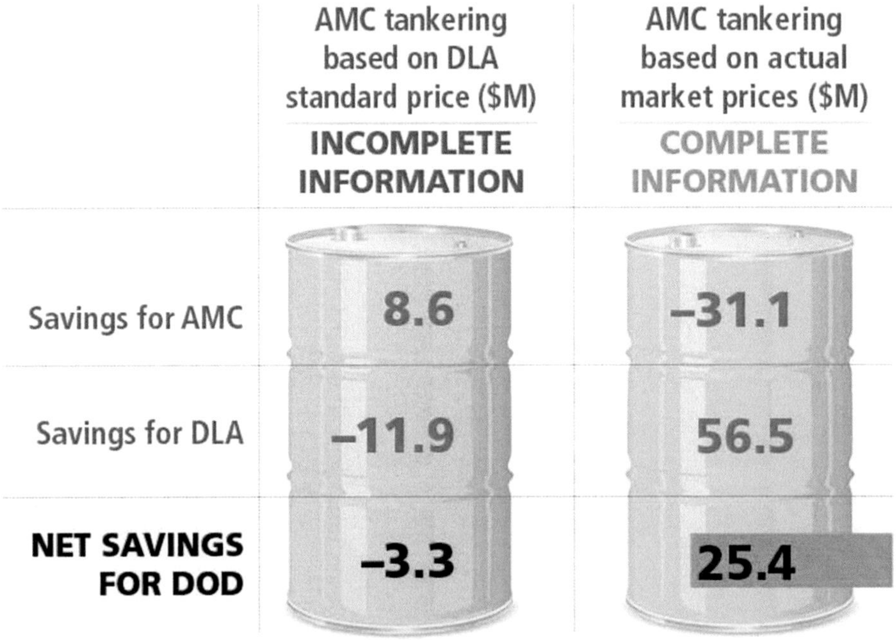

Given these findings, we can draw several conclusions and make some recommendations to AMC.

AMC can take advantage of differences in DLA standard prices. Savings from tankering are possible even without access to market fuel price data, if AMC takes advantage of differences among DLA standard prices. These savings are simple to achieve, because the prices are known, but they may occur only in the short run, as DLA has the option to make adjustments to its pricing schedule. In the long run, it makes sense for AMC and DLA to work together in developing a list of worldwide locations and the associated costs of fuel.

We recommend that USAF work with DLA to provide market price information to AMC to maximize the savings to DoD of peacetime tankering. Additionally, an internal compensation mechanism within DoD would be needed to shift some of the savings reaped by DLA to AMC to incentivize the command's participation. The need for a compensation mechanism would be eliminated if DLA simply charged market rates; however, this would introduce complications in the fuel budgeting process.

Tankering operations can be pursued to different extents and implemented in phases. A highly involved, complex tankering system might require optimization with extended visibility into future missions and aircraft assignments, real-time market prices for fuel at all locations, infrastructure investments to store and offload fuel, information technology investments to perform the linear programming techniques, and very high level coordination. However, as we have demonstrated, potential cost savings are possible even with less sophisticated tankering systems. Ideally, flight planners would be given a tool that they can use to input departure and arrival locations as well as mission payload, with a "lookup table" specifying the optimal fuel load. Development and trials of implementation could begin with basic tankering efforts. Given time and experience, a decision could be made on whether to invest further in more sophisticated implementations.[11]

[11] See Appendix D for possible avenues for research to inform such decisions.

Appendix A. DLA Standard Prices, April 2009–October 2012

(See the table on the following page.)

Table A.1. DLA Standard Prices, April 2009–October 2012

		DLA Standard Fuel Prices ($/gal)											
Location Type	**DLA Product Description**	**Apr-09**	**Sept-09**	**Oct-09**	**Jan-10**	**Jul-10**	**Oct-10**	**Jun-11**	**Oct-11**	**Jan-12**	**Jun-12**	**Jul-12**	**Oct-12**
DFSP	Turbine fuel, aviation, JP-8	1.44	2.13	2.78	2.82	2.34	3.03	3.95	3.95	3.82	3.60	2.31	3.73
With contract	Turbine fuel, aviation, Jet A (Into-Plane)	1.64	2.43	3.17	3.22	2.67	3.46	4.51	4.51	4.36	4.11	2.64	4.26
Without contract	Turbine fuel, aviation, Jet A (noncontract source at airport)	2.03	3.00	3.92	3.97	3.29	4.27	5.00	5.00	4.84	4.56	2.93	4.57
NATO	Turbine fuels, aviation, kerosene types, NATO F-34 (local purchase)	5.00	5.00	6.50	6.50	6.50	6.50	7.50	7.50	7.50	7.50	7.50	9.00

SOURCE: DLA (various dates).

Appendix B. Spot Price for Jet Fuel at 22 International Airports

The following table lists spot prices (excluding any sales or excise taxes) obtained from local fuel suppliers at 22 international locations between May 20 and May 31, 2013.

Table B.1. Spot Price for Jet Fuel at 22 International Airports

ICAO Code	Location	Into-Plane Rate for Jet A-1 Fuel ($/gal)
BKPR/LYPR	Pristina, Kosovo	5.12
EDDR	Saarbrücken, Germany	4.44
EDFH	Frankfurt–Hahn, Germany	3.39
EDSB	Rheinmünster, Germany	6.26
LEJR	Jerez, Spain	3.53
LICC	Catania, Italy	3.50
LIPH	Treviso, Italy	4.00
LIPZ	Venice, Italy	3.53
LTAF	Adana, Turkey	3.41
OAHR	Herat, Afghanistan	6.65
OAKB	Kabul, Afghanistan	6.22
OAKN	Kandahar, Afghanistan	6.50
OAMS	Mazar-i-Sharif, Afghanistan	6.46
OAUZ	Kunduz, Afghanistan	6.18
OMAA	Abu Dhabi, UAE	3.25
OOSA	Salalah, Oman	3.63
OPPS	Peshawar, Pakistan	3.33
PGUM	Barrigada and Tamuning, Guam	4.23
RJAA	Narita, Japan	3.19
RKSI	Incheon, South Korea	3.21
RKSS	Kimpo, South Korea	3.58
ROAH	Naha, Japan	3.23

Appendix C. Environmental Effect of Fuel Tankering

Fuel tankering and energy efficiency measures differ in their respective environmental effect. Energy efficiency and conservation measures reduce fuel costs by *decreasing* fuel consumption. Fuel tankering *increases* fuel consumption because of the increased burn penalty but decreases the overall fuel cost by displacing some of the fuel purchases from expensive to cheaper locations. Therefore, tankering savings come with increased aviation fuel emissions.

Table C.1 shows the extra fuel burn for the various cases simulated in this research. Results show that the burn penalty represents less than 2 percent of the total C-5, C-17, and C-130 fuel consumption when fuel offloading is not permitted but goes up to 6.5 percent of the total consumption when offloading is possible, i.e., more fuel is then tankered, hence a higher burn penalty. This has a negative environmental effect that would need to be considered further.

Table C.1. Extra Fuel Burn Compared to Total Fuel Consumption and Total Tankering Savings for DoD

		Wartime Scenario			Peacetime Scenario		
Level of Price Information	**Offloading of Tankered Fuel**	Extra fuel burn (MG)	% of total fuel consumption on AMC aircraft considered	DoD savings per unit of extra fuel burn ($M/MG)	Extra fuel burn (MG)	% of total fuel consumption on AMC aircraft considered	DoD savings per unit of extra fuel burn ($M/MG)
AMC tankers based on DLA standard prices (incomplete information)	No offloading	5.54	0.68	14.23	1.90	0.27	-1.76
	Offloading from tank only	21.59	2.59	8.69	7.38	1.05	-2.61
	Offloading from tank and cargo hold	32.85	3.89	7.43	11.36	1.61	-2.63
AMC tankers based on actual market prices (complete information)	No offloading	12.90	1.57	8.54	7.68	1.09	3.31
	Offloading from tank only	37.77	4.45	11.18	18.66	2.62	4.09
	Offloading from tank and cargo hold	56.39	6.50	10.90	28.23	3.91	4.06

Appendix D. Future Research Directions

Modeling of Mission Profiles

Additional research is needed to improve the way mission profiles are modeled. In this research, we assumed for simplicity that the information on the downstream mission fuel requirements for any given aircraft was simply the average profile of the next mission leg (Chapter 2).

A possible improvement would be to develop an optimization model that optimally allocates over a rolling period of seven to ten days the amounts of fuel purchased across the various locations visited and the amounts of fuel tankered across the various mission legs.

The development of such a tool—similar to the software used by commercial air carriers—would serve both planning and operational purposes. It would be descriptive (quantifying the savings generated by tankering fuel), prescriptive (providing concrete guidance to pilots), and normative (demonstrating how tankering decisions should be made across the board to maximize tankering savings for DoD).

Feasibility of Fuel Offloading

Further research might also examine the feasibility of fuel offloading, which this work showed is related to a significant increase in tankering savings. Our preliminary assessment suggests that the effect of fuel offloading operations on the logistics services on the ground—and on the use and availability of fuel trucks in particular—constitutes a significant barrier to implementation.

Future research should examine in greater detail the size, availability rates, and ownership of the existing truck fleets serving AMC aircraft at the various worldwide locations and explore the possibility of expanding these fleets to enable fuel offloading operations if it makes sense economically. Finally, in this work, we assumed for simplicity that the demand for jet fuel was not a limiting factor. Future research on the feasibility of fuel offloading should test the validity of this assumption across the worldwide locations considered using historical data.

Externalities and Compensation Mechanisms Within DoD

A third area for future research would focus on the possible need for compensation mechanisms within DoD as well as possible externalities that may result from an AMC tankering program. Our results uncovered the need for internal compensation mechanisms in several cases where AMC was penalized for tankering fuel even as it was generating savings for DoD as a whole. For example, when AMC and DLA cooperate under the peacetime scenario, fuel tankering generates $25 million in annual savings for DoD but incurs $31 million in annual losses to AMC. This result is a net $56 million annual windfall for DLA in the form of savings and additional revenues (Table 4.5). A compensation mechanism might involve redistributing part of this additional income toward AMC.

Externalities resulting from tankering fuel are a related area for potential future research. In particular, second-order effects on other DoD commands should be carefully studied (Figure D.1). The first case that we examined in this research—the wartime scenario with incomplete information—illustrates this point. Under this first case, we found that tankering fuel on AMC aircraft would lead to $150 million in annual savings for AMC and almost $79 million for DoD (Table 4.3). In this scenario, DLA faces a budget shortfall of $71 million annually. A natural remedy could involve increasing DLA's standard fuel prices and charging AMC higher rates. However, since the standard prices are the same across all the U.S. Armed Forces, this price increase would artificially affect other commands that also buy aviation fuel from local suppliers and are charged at the DLA rates. In other words, other commands could be affected financially by AMC tankering efforts, all other things being equal. This simple example shows the need for further analysis in this area.

Figure D.1. Jet Fuel Procurement Process

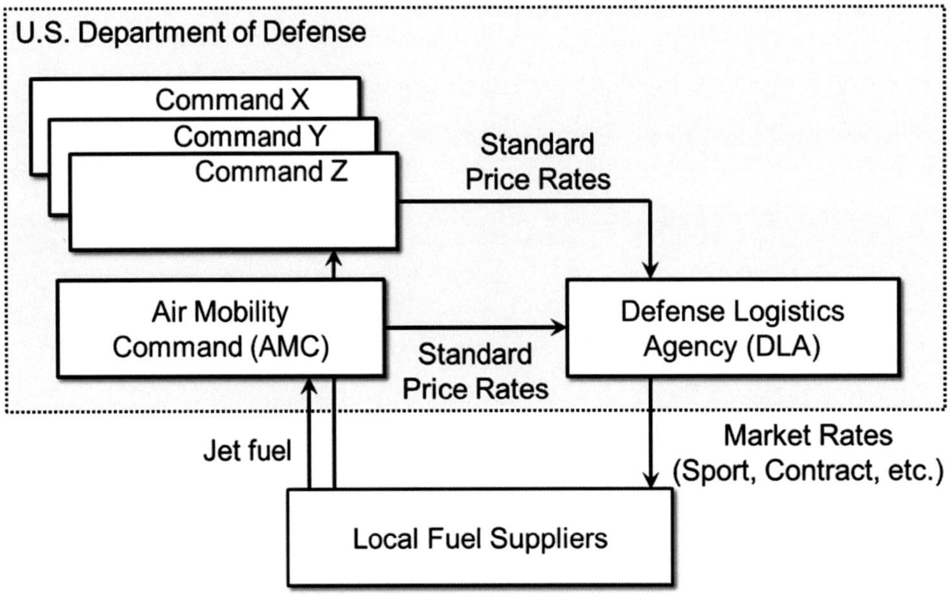

SOURCE: Authors.

References

Abdelghany, Khaled, Ahmed Abdelghany, and Sidhartha Raina, "A Model for the Airlines' Fuel Management Strategies," *Journal of Air Transport Management*, Vol. 11, No. 4, 2005, pp. 199–206.

Airbus, *Getting to Grips with Fuel Economy*, Flight Operations Support & Line Assistance, July 2004.

Bednarz, Sean, Anthony D. Rosello, Shane Tierney, David Cox, Steven C. Isley, Michael Kennedy, Chuck Stelzner and Fred Timson, *Modernizing the Mobility Air Force for Tomorrow's Air Traffic Management System*, Santa Monica, Calif.: RAND Corporation, MG-1194-AF, 2012. As of September 29, 2014: http:www.rand.org/pubs/monographs/MG1194.html

Congressional Research Service (CRS), *Department of Defense Energy Initiatives: Background and Issues for Congress*, 2012.

Darnell, D. Wayne, and Carolyn Loflin, "National Airlines Fuel Management and Allocation Model," *Interfaces*, Vol. 7, No. 2, 1977, pp. 1–16.

Defense Logistics Agency (DLA), "Into Plane Contract Information System," undated-a. As of September 29, 2014: https://ports.energy.dla.mil/ip_cis/ipcis.htm

———, *DLA Energy Standard Prices*, undated-b. As of September 1, 2013: http://www.energy.dla.mil/customers/standard_prices/Pages/default.aspx

———, *Defense Logistics Agency Energy Factbook Fiscal Year 2012*, Fort Belvoir, Va., undated-c. As of September 29, 2014 : http://www.energy.dla.mil/library/Pages/Publications.aspx

Federal Aviation Administration, *Aircraft Fuel Control*, AC 20-43C, October 20, 1976.

———, *Aviation Fuel—Identification and Fueling Procedures*, AC 20-106, Appendix B, April 1978.

———, *Water in Aviation Fuel*, AC 20-125, November 12, 1985.

Fregnani, Guerreiro, José Alexandre Tavares, Carlos Müller, and Anderson Ribeiro Correia, "A Fuel Tankering Model Applied to a Domestic Airline Network," *Journal of Advanced Transportation*, Vol. 47, No. 4, 2013, pp. 386–398.

Fritz, Oliver, *Operational Energy Considerations*, Directorate of Strategic Planning, Headquarters U.S. Air Force, June 14, 2010, p. 15.

Health, E., *Air Force Jp-8 Fuel Distribution System: A Statistical Analysis to Determine Where and When to Sample*, March 2005.

Kheraie, Ali Zockaie, and Hani S. Mahmassani, "Leveraging Fuel Cost Differences in Aircraft Routing by Considering Fuel Ferrying Strategies," *Transportation Research Record: Journal of the Transportation Research Board*, Vol. 2300, No. 1, 2012, pp. 139–146.

Langton, R., C. Clark, M. Hewitt, L. Richards, *Aircraft Fuel Systems*, Wiley, 2009.

Lesinski III, Walter J., *Tankering Fuel: A Cost Saving Initiative*, June 2011.

Mouton, Christopher A., David T. Orletsky, Michael Kennedy, and Fred Timson, *Reducing Long-Term Costs While Preserving a Robust Strategic Airlift Fleet: Options for the Current Fleet and Next-Generation Aircraft*, Santa Monica, Calif.: RAND Corporation, MG-1238-AF, 2013. As of September 29, 2014: http://www.rand.org/pubs/monographs/MG1238.html

Mouton, Christopher A., Jim D. Powers, Daniel M. Romano, Christopher Guo, Sean Bednarz, and Caolionn O'Connell, *Fuel Reduction for the Mobility Air Forces,* Santa Monica, Calif.: RAND Corporation, RR-757-PR, 2014a. As of September 29, 2014: http://www.rand.org/pubs/research_reports/RR757.html

Mouton, Christopher A., Jim D. Powers, Daniel M. Romano, Christopher Guo, Sean Bednarz, and Caolionn O'Connell, *Fuel Reduction for the Mobility Air Forces: Executive Summary,* Santa Monica, Calif.: RAND Corporation, RR-757/1-PR, 2014b. As of September 29, 2014: http://www.rand.org/pubs/research_reports/RR757z1.html

Nash, Barry, "A Simplified Alternative to Current Airline Fuel Allocation Models," *Interfaces*, Vol. 11, No. 1, 1981, p. 1.

Rampmaster, *Rampmaster 17,500 Gallon WD Standard Specification*, undated. As of September 1, 2013: http://www.rampmasters.com/pdf/rampmaster-17500-spec.pdf

Rosello, Anthony D., Sean G. Bednarz, Michael Kennedy, Chuck Stelzner, F. S. Timson, and David T. Orletsky, *Assessing the Cost-Effectiveness of Modernizing the KC-10 to Meet Global Air Traffic Management Mandates*, Santa Monica, Calif.: RAND Corporation, MG-901-AF, 2009. As of September 29, 2014: http://www.rand.org/pubs/monographs/MG901.html

Rosello, Anthony D., Sean G. Bednarz, David T. Orletsky, Michael Kennedy, F. S. Timson, Chuck Stelzner, and Katherine M. Calef, *Upgrading the Extender: Which Options Are Cost-Effective for Modernizing the KC-10?* Santa Monica, Calif.: RAND Corporation, TR-901-AF, 2011. As of September 29, 2014: http://www.rand.org/pubs/technical_reports/TR901.html

Singh, Vedant, and Somesh K. Sharma, "Evolving Base for the Fuel Consumption Optimization in Indian Air Transport: Application of Structural Equation Modeling," *European Transport Research Review*, 2014, pp. 1–18.

Stroup, John S., and Richard D. Wollmer, "A Fuel Management Model for the Airline Industry," *Operations Research*, Vol. 2, 1992, p. 229.

Titan Aviation, *40,000L Semi Trailer Refueller*, undated. As of September 1, 2013: http://brochures.titan-aviation.com/titan-aviation-40000l-semitrailer-refueller-en.pdf

U.S. Air Force, *Fuels Logistics Planning*, AFPAM 23-221, December 2006.

———, *AF Energy Plan 2010*, AFD-091208-027, Washington, D.C., SAF IE, 2010a.

———, *AF Aviation Operation Energy Plan 2010*, AFD-091208-026, Washington, D.C., SAF IE, 2010b.

———, Florex, Pascual, *Joint Base McGuire Selected to Test Drive New Fuel Truck*, Joint Base McGuire-Dix-Lakehurst Public Affairs Press Release, May 5, 2010c. As of September 1, 2013: http://www.af.mil/News/ArticleDisplay/tabid/223/Article/116775/joint-base-mcguire-selected-to-test-drive-new-fuel-truck.aspx

U.S. Air Force, *Aircraft Fuel Servicing with R-9, R-11, and Commercial Fuel Servicing Trucks*, TO 00-25-172CL-4, May 2012.

U.S. Air Force, *Ground Servicing of Aircraft and Static Grounding/Bonding*, TO 00-25-172, August 2013.

U.S. Air Force, *Mobility Air Forces Cost Avoidance Tankering (MAFCAT)*, USTRANSCOM CBDS Working Group, January 2014.

U.S. Army, *Concepts and Equipment of Petroleum Operations*, FM 10-67-1, 1998.

U.S. Army, *Forward Arming and Refueling Point, Tactics, Techniques, and Procedures*, FM 3-04.104, 2006.

U.S. Navy, "Operational and Training Facilities," *Facility Planning for Navy and Marine Corps Shore Installations*, UFC 2-000-05N, 2005.